E-Learning's Enduring Legacy

Enhancing Online Education Through Technology

Bennett Pendleton

AF719888

Contents

1 Introduction to E-Learning

E-learning's origins lie in correspondence courses and early forms of distance education, utilizing print materials and radio. The advent of the internet revolutionized this approach, creating a dynamic, interactive learning environment accessible globally. Today, e-learning encompasses a vast spectrum of methods, from asynchronous learning modules to live virtual classrooms employing advanced multimedia and collaborative tools. This evolution continues, driven by technological innovation and a growing demand for flexible and accessible educational opportunities.

This shift to online learning offers significant advantages. Cost-effectiveness is a key benefit; reduced overhead compared to traditional institutions translates to lower tuition fees for students. Furthermore, geographical barriers vanish. E-learning transcends physical limitations, enabling access to higher education for individuals in remote areas or with mobility challenges. The flexibility inherent in online learning caters to diverse learning styles and schedules, accommodating working professionals, parents, and students with other commitments. The availability of resources, ranging from digital libraries to interactive simulations, expands learning opportunities beyond the confines of a physical classroom.

Designing effective e-learning courses requires careful consideration of several key factors. First, course structure must be meticulously planned, ensuring a clear, logical progression of content. Effective navigation is essential; a user-friendly interface that is easy to follow and intuitive is critical for positive student experience. Well-organized modules, coupled with clear learning objectives and assessments, provide a structured learning journey. Regular updates and maintenance are essential to prevent outdated or broken links, ensuring the course remains engaging and functional. Finally, accessibility should be a top priority in the design process.

The integration of technology is crucial in creating dynamic and engaging virtual learning environments. Learning management systems (LMS) provide a central hub for course materials, communication, and assessment. These platforms facilitate interaction between instructors and students, fostering a sense of community. Multimedia resources, such as videos, simulations, and interactive exercises, enhance engagement and cater to different learning preferences. Selecting appropriate technologies depends on several factors; cost-effectiveness, ease of use, and compatibility with existing systems must be balanced against the desired level of functionality and pedagogical goals.

Effective e-learning isn't simply about delivering content; it's about fostering meaningful engagement. Active learning strategies, like discussions forums, interactive quizzes, and collaborative projects, encourage active participation. Collaborative learning techniques further enhance the educational experience by promoting peer interaction, skill-building, and knowledge sharing. Gamification, the integration of game design elements, adds an element of fun and competition, motivating students to actively participate and achieve learning goals. Regular, constructive feedback is crucial; personalized feedback tailored to individual student needs promotes learning and helps students identify areas for improvement.

Finally, the success of any e-learning initiative hinges on effective communication and dedicated student support. Clear communication channels should be established, ensuring timely responses to student inquiries. Regular interaction through announcements, online forums, or virtual office hours maintains a sense of community and fosters a supportive learning environment. Proactive student support extends beyond responding to questions; it involves providing resources, guidance, and encouragement to help students succeed. Mentorship programs or peer support networks can build a strong sense of community within the online learning environment, crucial for student engagement and overall success. Ultimately, creating a supportive and engaging environment empowers learners to reach their full potential.

1.1 The Evolution of E-Learning

Early forms of e-learning, like correspondence courses, relied heavily on printed materials and mailed assignments. The advent of the internet dramatically altered this landscape, ushering in a new era of asynchronous learning. Platforms like Blackboard and Moodle emerged, offering a more interactive experience, allowing for online discussions and the sharing of digital resources. This shift marked a significant leap forward, transforming the accessibility and reach of education. However, the early days were also hampered by technical limitations and a lack of pedagogical understanding of how to best leverage the technology. Early adopters often struggled to replicate the dynamism of a traditional classroom.

The rise of broadband internet access in the late 1990s and early 2000s provided the necessary infrastructure for richer online learning experiences. Multimedia content, including videos, simulations, and interactive exercises, became more commonplace. The integration of these elements significantly enhanced engagement and knowledge retention. Furthermore, the emergence of learning management systems (LMS) provided a centralized hub for all course materials and activities, streamlining the learning process for both instructors and students. This period witnessed a significant increase in the adoption of e-learning across various sectors, from higher education to corporate training.

The introduction of Web 2.0 technologies, such as social media platforms and collaborative tools, marked another major turning point. Online learning communities fostered a sense of connection and belonging among students, often missing in traditional distance learning. Tools like wikis and blogs encouraged participation and peer-to-peer learning, promoting a more active learning experience. The rise of mobile devices also broadened accessibility even further, allowing students to access course materials and engage in learning activities anytime, anywhere. This era brought the potential of personalized learning to the forefront.

Modern e-learning platforms now incorporate sophisticated features designed to personalize the learning experience. Adaptive learning technologies leverage data analytics to tailor content and pacing to individual student needs and learning styles. Artificial intelligence-driven systems are increasingly used to provide personalized feedback, identify knowledge gaps, and recommend relevant resources. This targeted approach maximizes learning outcomes by focusing on each student's unique strengths and weaknesses. Consequently, the role of the educator has evolved, shifting from a simple content provider to a facilitator of learning, mentor, and guide.

Current trends point toward an even more immersive and personalized future. Virtual reality (VR) and augmented reality (AR) technologies offer opportunities to create highly engaging and interactive learning experiences. These technologies can transport students to virtual environments, allowing them to interact with complex concepts in a tangible way. The development of personalized learning pathways, using algorithms and analytics to design tailored education plans, promises to further enhance learning effectiveness and efficiency. Moreover, the increasing integration of gamification techniques, such as points, badges, and leaderboards, can boost motivation and enhance overall engagement. This evolution continues apace, constantly refining and developing in response to pedagogical advancement and technological innovation.

The evolution of e-learning has been nothing short of transformative. From its humble beginnings in correspondence courses to the sophisticated platforms of today, e-learning has continually adapted and improved. This trajectory demonstrates a continuing commitment to enhance accessibility, engagement, and the overall quality of the learning experience. The future promises further advancements, pushing the boundaries of what's possible in online education. The technology remains merely a tool; the success depends on thoughtful pedagogical approaches and a commitment to accessibility and inclusivity for all learners. E-learning's enduring legacy lies in its ability to constantly evolve, meeting the changing demands of learners and educators alike. The journey has been remarkable, and the destination promises even greater possibilities.

1.2 Benefits of Online Education

Online education offers unparalleled flexibility. Students can learn at their own pace, fitting study around existing commitments like work or family responsibilities. This asynchronous learning style empowers individuals to manage their time effectively, enhancing their overall learning experience and reducing the stress often associated with traditional classroom schedules. The convenience is undeniable; accessing course materials and participating in discussions anytime, anywhere, with a reliable internet connection, truly democratizes education.

Unlike traditional classrooms limited by geographical constraints, online learning transcends borders. Students from anywhere in the world can access educational resources and interact with instructors and peers, fostering global collaboration and intercultural understanding. This expanded reach opens doors to learning opportunities that would otherwise be inaccessible, leveling the playing field and promoting educational equity on an international scale. This global reach contributes significantly to a more diverse and inclusive learning environment.

The cost-effectiveness of online education is a significant benefit. Reduced commuting expenses, eliminating the need for on-campus housing, and often lower tuition fees compared to traditional institutions make online learning a financially viable option for many. These savings can be substantial, especially for students facing financial burdens, enabling them to pursue higher education without incurring crippling debt. This affordability promotes wider access to education, a crucial aspect of personal and societal development.

Online learning frequently employs innovative teaching methodologies. Interactive simulations, virtual labs, and multimedia resources create engaging and immersive learning experiences, moving beyond the limitations of traditional lecture formats. These dynamic tools enhance knowledge retention and comprehension, catering to diverse learning styles and preferences. The integration of these technologies ensures that learning is not just passive, but active and participatory.

Furthermore, online education facilitates personalized learning experiences. Adaptive learning platforms and personalized feedback mechanisms allow instructors to tailor their instruction to the individual needs and pace of each student. This targeted approach maximizes individual learning outcomes, ensuring that every student receives the support they require to succeed. This personalized learning pathway ultimately promotes greater student engagement and success.

The accessibility of online education extends beyond just geographic limitations. Online courses frequently provide materials in accessible formats, catering to students with disabilities. This inclusive approach ensures equal opportunities for all learners, regardless of their physical or cognitive abilities. The focus on accessibility promotes a more equitable and inclusive learning environment for all students, regardless of their individual circumstances.

The immediate feedback mechanisms inherent in many online learning environments are incredibly beneficial. Online quizzes, assignments, and discussions provide students with prompt feedback on their progress, allowing them to identify areas where they need improvement and adjust their learning strategies accordingly. This immediate feedback loop is crucial for effective learning and continuous improvement. It empowers students to take control of their learning journey and actively address any challenges they encounter. Moreover, online learning platforms often incorporate robust communication tools. Discussion forums, email, and video conferencing facilitate continuous interaction between students and instructors, fostering a sense of community and support. This constant communication ensures that students feel connected to their learning environment and can readily seek help when needed. The sense of community is crucial for online learning success, combating the feelings of isolation that can sometimes accompany online education.

Finally, the digital nature of online education lends itself to readily available and accessible resources. A wealth of information, from scholarly articles to interactive simulations, is readily available through online learning platforms and the internet. This readily available access to information expands the scope of learning beyond the confines of any single textbook or lecture, encouraging students to explore topics in depth and develop their research skills. This access to unlimited resources is a definite advantage of online learning.

In conclusion, the benefits of online education are numerous and profound. From its flexibility and cost-effectiveness to its innovative teaching methodologies and inclusive approach, online learning offers a transformative educational experience. It's not merely a substitute for traditional education, but a powerful complementary modality that enhances access, promotes personalized learning, and empowers learners in ways that were previously unimaginable. The future of education is undoubtedly shaped by the enduring legacy of e-learning.

2 Designing Effective Virtual Classrooms

Consider the virtual classroom not as a mere digital replica of a physical one, but as a dynamic, interactive space demanding careful design. Its success hinges on intuitive navigation and a clear, visually appealing structure. Think of modules clearly labeled and easily accessible, a consistent layout reducing cognitive load, and a streamlined interface minimizing distractions. The goal isn't just functionality; it's seamless engagement, encouraging active participation rather than passive consumption.

Effective use of technology is pivotal; it's not about quantity, but strategic integration. Overloading students with multiple platforms will only confuse and frustrate. Select tools that enhance learning, not merely replace traditional methods. Consider interactive whiteboards for collaborative brainstorming, video conferencing for real-time discussions, and learning management systems for streamlined communication and resource management. Regular technology checks are essential, minimizing disruption and ensuring a smooth learning journey for everyone.

Interactive learning tools are more than mere add-ons; they are the heart of an engaging virtual classroom. Think beyond simple quizzes; incorporate simulations, virtual labs, and interactive case studies mirroring real-world scenarios. These tools foster active learning, encourage critical thinking, and provide opportunities for practical application of concepts. Remember, the aim is to make learning enjoyable and relevant, stimulating interest and boosting knowledge retention. Carefully select these tools based on your specific learning goals and student needs, ensuring they complement the overall learning design. Poorly integrated technology can easily distract from learning; thoughtfully-chosen tools, however, significantly enhance it.

The layout of the virtual classroom should be meticulously planned, just as a well-designed physical classroom maximizes space and accessibility. Consider dividing the space into zones for different activities – a dedicated area for announcements, a space for collaborative projects, and a section for individual study. Think of it as a well-organized digital workspace, intuitive to navigate and free from clutter. A disorganized classroom mirrors a disorganized mind, and can significantly impair the learning experience. Ensure that all elements are clearly positioned and easily accessible, preventing frustration and facilitating smooth engagement.

Beyond the technical aspects, fostering a sense of community is crucial. Encourage interaction through discussion forums, group projects, and virtual office hours. Regular communication, offering both scheduled and unscheduled opportunities for interaction, keeps students feeling connected and supported. This sense of belonging is especially important in the virtual environment, counteracting feelings of isolation that can hinder learning progress. A vibrant online community significantly improves student engagement and the overall learning atmosphere.

Don't underestimate the power of visual elements. A visually stimulating classroom captures attention and enhances comprehension. Use a consistent color scheme, high-quality images and videos, and clear typography. Avoid overwhelming students with excessive visual clutter, however, balancing visual appeal with functional simplicity. A well-designed visual environment enhances the learning experience, making it more memorable and engaging. Remember, even the smallest design elements, when carefully considered, can profoundly impact the classroom's overall feel.

Accessibility for all learners is not optional; it's a necessity. Ensure your classroom design caters to diverse needs, considering students with disabilities. Use clear and concise language, provide alternative formats for materials, and use captions for videos. Employ assistive technologies where necessary and ensure compatibility with screen readers and other assistive devices. Accessibility isn't merely a compliance issue; it is a crucial aspect of creating a truly inclusive and equitable learning environment, beneficial for all students. Designing an inclusive virtual classroom ensures that everyone can participate fully and achieve their learning goals.

Finally, continuous evaluation and refinement are key. Gather feedback regularly to identify areas for improvement and adapt your virtual classroom accordingly. Regular monitoring of student engagement, coupled with their feedback, is crucial to the long-term success of any online course. This iterative process of design, implementation, and feedback enables you to create a virtual classroom that is truly effective and responsive to the needs of your learners. Continuously improving your virtual classroom ensures that it remains a dynamic and effective learning space for all.

2.1 Classroom Structure & Layout

The layout of your virtual classroom significantly impacts student engagement and learning outcomes. Think of it as the stage for your online performance – a poorly designed stage can hinder even the most talented actor. A well-structured virtual classroom, however, creates a natural flow, guiding learners through the material effortlessly. Consider using a modular design, breaking down complex topics into smaller, manageable chunks. This segmented approach prevents information overload, allowing for better comprehension and retention. Organize your modules logically, progressing from foundational concepts to more advanced material. Clearly defined sections will reduce student confusion and frustration, leading to a more positive learning experience.

Navigation is key. Imagine trying to find your way through a maze without a map; frustrating, right? Your virtual classroom needs intuitive navigation, allowing students to easily access materials, assignments, and announcements. A well-designed menu, clear visual cues, and a consistent layout are essential. Ensure that all resources are easily accessible from a central location, reducing the time students spend searching and maximizing their learning time. A logical structure translates to a more efficient and effective learning environment. Don't forget to regularly review your navigation design, making adjustments as needed based on student feedback and evolving needs.

Visual appeal matters just as much as functionality. A visually appealing classroom is more inviting and engaging. Incorporate high-quality images, videos, and interactive elements to break up text-heavy content and maintain student interest. Consistent branding and a cohesive aesthetic create a professional and welcoming atmosphere. Think of it as setting the mood; a cluttered, chaotic environment is distracting, whereas a clean, organized space promotes focus and concentration. Use color strategically to highlight key information and create visual interest, while maintaining readability and accessibility for all learners. A visually appealing virtual classroom contributes significantly to a positive learning experience.

Consider the use of different learning spaces within your virtual classroom. Designate specific areas for different activities – a discussion forum for collaborative work, a resource library for supplemental materials, and an announcements section for important updates. This compartmentalization helps students quickly locate necessary information and reduces confusion. Clearly label each section to ensure easy navigation and understanding. Implementing separate learning spaces fosters a more organized and efficient online learning environment, improving the overall learning experience for students. Remember that the organization of your virtual classroom directly impacts student engagement and effectiveness of learning.

Remember accessibility. Your virtual classroom should be accessible to all students, regardless of their abilities or technological limitations. Use clear fonts, sufficient contrast between text and background, and provide alternative text for images. Consider using closed captioning for videos and ensuring that all materials are compatible with assistive technologies. Accessibility is not just a matter of compliance but a matter of equity and inclusivity; every student deserves equal access to the learning materials and opportunities. Prioritizing accessibility creates a more inclusive environment for everyone. A meticulously designed virtual classroom, considering accessibility from the outset, creates a welcoming and inclusive atmosphere that serves all learners effectively. Prioritizing this from the planning stage reduces potential hurdles later, optimizing the learning process for every participant.

The use of technology plays a critical role in effective classroom structure. Choosing the right tools for interaction, collaboration, and assessment is crucial. Tools facilitating group work and discussion are integral for active learning, promoting interaction between students. Consider integrated video conferencing, interactive whiteboards, or collaborative document editing tools, all designed to enhance communication and teamwork. These technological choices significantly impact the dynamic of the classroom environment and should be carefully evaluated to support the overall learning objectives. Strategic selection ensures a positive learning experience, enhancing student interaction and collaboration.

Regularly evaluate and update your virtual classroom. Gather feedback from your students and analyze learning outcomes to identify areas for improvement. Are students struggling to navigate the classroom? Is the layout confusing or inefficient? Regularly updating your virtual classroom ensures it remains functional, engaging, and relevant. This iterative process demonstrates your commitment to providing an optimal learning environment. Continuously refining your virtual classroom in response to student feedback shows commitment to quality and improves overall student satisfaction and learning outcomes. Remember, the classroom should constantly evolve to reflect the needs of your students and the advancement in educational technology.

2.2 Utilizing Technology Effectively

Effective technology integration isn't about simply adding gadgets to your virtual classroom; it's about strategically choosing tools that enhance learning. Consider your learning objectives: Do you need interactive simulations for complex concepts, collaborative platforms for group projects, or automated assessment tools for efficient feedback? Choosing the right tools directly correlates to student engagement and understanding.

Start small. Don't overwhelm yourself or your students by introducing a multitude of new technologies all at once. Begin with one or two carefully selected tools, mastering their functionalities before integrating others. This phased approach allows for smoother transitions and avoids potential technical difficulties that might hinder the learning process. Remember that effective technology integration is an iterative process. Regularly evaluate the effectiveness of your chosen tools; are they achieving their intended purpose? Are students finding them user-friendly and beneficial to their learning? Be prepared to adapt and adjust your approach based on feedback and observed outcomes.

The most advanced technology won't compensate for poor pedagogical design. Your choice of technology should directly support your teaching methodology and overall learning goals. For instance, if your goal is to foster collaborative learning, choose tools that facilitate group work and communication. Likewise, if your focus is on individual learning, choose tools that provide personalized feedback and adaptive learning paths.

Effective communication is critical in online learning. Regular announcements, timely responses to student queries, and clear instructions are essential. Utilize various communication channels—forums, email, video conferencing—to cater to diverse student preferences and needs. Remember, effective communication fosters a sense of community and belonging, crucial elements in a successful online learning environment.

Remember that technology serves as a facilitator, not a replacement, for effective teaching. Don't rely solely on technology to engage your students. Incorporate active learning strategies, such as discussions, quizzes, and assignments, to keep them actively involved in the learning process. This blended approach ensures that technology complements, rather than overshadows, the core teaching and learning experience. Strive for a balanced approach that leverages technology's strengths while maintaining the human connection integral to effective teaching.

Accessibility is paramount. Ensure all chosen technologies are accessible to students with disabilities. Consider using captioning for videos, providing transcripts for audio materials, and offering alternative formats for visual content. This inclusivity ensures that all students have equitable access to the learning materials and opportunities.

Data analytics offers valuable insights into student performance and engagement. Many Learning Management Systems (LMS) provide data on student activity, assignment completion rates, and quiz scores. Analyzing this data can help identify areas where students are struggling and adjust your teaching strategies accordingly. It allows for personalized interventions and ensures that learning remains effective for all participants. Remember, the goal is not simply to collect data but to use it to inform teaching practices and improve learning outcomes.

Finally, continuous professional development is key to staying current with technological advancements and best practices. Explore online resources, attend workshops, and engage with colleagues to learn about new technologies and innovative teaching strategies. Staying informed allows you to consistently improve your online teaching and provide an engaging and effective learning experience for your students. Embrace lifelong learning to continually refine your use of technology in education.

2.3 Interactive Learning Tools

Interactive simulations offer a powerful way to engage students actively. Imagine a virtual lab where students dissect a frog without the mess, or a historical reenactment where they participate in pivotal moments. These immersive experiences transcend passive learning, fostering deeper understanding and retention through hands-on exploration. Properly designed simulations can cater to diverse learning styles, allowing visual, auditory, and kinesthetic learners to connect with the material in a personalized way. The key is to select simulations that align with learning objectives and provide meaningful feedback. Careful integration is crucial; a poorly chosen or poorly implemented simulation can distract from learning rather than enhancing it.

Interactive whiteboards and collaborative annotation tools provide a dynamic platform for real-time interaction. These tools transform lectures into active discussions, facilitating student participation and collaboration. Imagine students annotating a shared document during a lecture, contributing their thoughts and questions directly onto the material. This promotes critical thinking and shared understanding. The instructor can easily guide the discussion, address misconceptions, and assess comprehension in real time. Beyond direct lecture enhancement, these tools can also be invaluable for group projects and collaborative assignments, promoting teamwork and communication skills. Selecting the right whiteboard or annotation tool will depend on the specific needs of your course and the technical capabilities of your students, factors to carefully consider for successful implementation.

Polls, quizzes, and interactive games are not merely distractions; they're valuable assessment and engagement tools. Short, frequent polls can gauge student understanding throughout a lecture, providing immediate feedback and opportunities for clarification. Interactive quizzes can test knowledge and reinforce learning in a fun and engaging way. Well-designed games can simulate real-world scenarios, encouraging problem-solving and critical thinking skills. It's imperative to use these tools judiciously to avoid overwhelming students or detracting from the core content. The goal is to enhance learning, not replace it with trivial activities. The careful design of these activities and the subsequent analysis of student responses can be a powerful way to tailor instruction and improve student outcomes.

Discussion forums and online chat features foster a sense of community and facilitate peer-to-peer learning. These tools allow for asynchronous communication, enabling students to engage with each other and the instructor outside of scheduled class times. Well-moderated discussion forums can encourage critical thinking, debate, and collaborative problem-solving. Online chat features can facilitate quick questions and answers, creating a sense of immediacy and responsiveness. However, managing these tools effectively requires clear guidelines and active moderation to ensure productive and respectful interactions. Students should be encouraged to engage thoughtfully and respectfully with one another, understanding the importance of contributing meaningfully to the discussion. Neglecting these guidelines risks creating a chaotic, unmanageable environment, hindering learning rather than supporting it.

Virtual field trips and immersive experiences are increasingly accessible and offer unparalleled learning opportunities. These can range from virtual tours of museums and historical sites to simulations of scientific experiments or even simulations of business contexts. Such immersive experiences provide students with a richness of context and information, bringing the learning to life in a way that traditional methods cannot. Careful selection of these experiences ensures relevance to the course materials and learning objectives. By integrating such advanced resources effectively, instructors can foster a more engaging and memorable learning environment. Moreover, the potential for virtual collaboration and group work within these immersive environments creates a unique opportunity for teamwork and shared learning. This creates an opportunity to foster not only subject matter mastery but also essential collaborative skills.

3 Creating Engaging Learning Experiences

Designing truly engaging e-learning experiences hinges on understanding your students. Think about their learning styles, their preferred methods of interaction, and their individual needs. Don't just lecture; create an active learning environment. Consider implementing activities that promote interaction and collaboration, fostering a sense of community even in a virtual setting.

For example, incorporate interactive quizzes or polls throughout the modules. These brief assessments not only check for understanding but also keep students actively engaged. They provide immediate feedback, helping students identify knowledge gaps and reinforce learning in real-time, preventing passive consumption of content. This dynamic approach encourages participation and transforms learning from a one-way street into a lively exchange.

Collaborating effectively requires more than just assigning group projects. Structure the collaborative activities carefully. Provide clear guidelines, define roles, and establish communication channels. Employ tools that facilitate seamless collaboration, such as shared documents, online whiteboards, or collaborative project management software. Consider asynchronous activities for students in different time zones.

Gamification, when implemented thoughtfully, can dramatically enhance engagement. Introduce points, badges, leaderboards, and challenges to motivate students and make learning more enjoyable. However, it's crucial to align the game mechanics with learning objectives, avoiding superficial gamification that distracts from core concepts. The game should enhance the learning process, not replace it.

Assessment isn't just about grading; it's about providing valuable feedback. Use a variety of assessment methods, incorporating formative assessments throughout the course to gauge understanding and provide timely feedback. This allows for adjustments in teaching strategies and provides students with opportunities to improve. Summative assessments, such as exams or projects, should align with learning objectives and provide a comprehensive evaluation of student learning.

Personalized learning paths cater to individual student needs and preferences. Offer a variety of learning materials, such as videos, podcasts, readings, and interactive simulations. Allow students to choose the materials that best suit their learning styles. Use adaptive learning platforms that adjust the difficulty and content based on individual student progress. Personalization isn't simply about choice; it's about recognizing and responding to individual learning journeys.

Consider incorporating real-world case studies, relevant examples, and relatable scenarios to make the material more engaging. Students learn better when they can connect what they're learning to their own lives and experiences. Avoid abstract concepts; instead, ground them in practical applications. Incorporate multimedia elements, such as videos and images, to cater to various learning styles and enhance understanding. The goal is to create a dynamic, engaging, and accessible learning environment.

Furthermore, foster a sense of community by encouraging student interaction. Use discussion forums, online chats, and virtual office hours to facilitate communication and create opportunities for peer-to-peer learning. Respond promptly to student queries and provide personalized support. Create a welcoming and inclusive online environment where students feel comfortable participating and asking questions. Remember, effective communication is key to a successful learning experience.

Effective communication goes beyond simply delivering lectures. Use a variety of communication tools and methods, tailoring your communication style to the specific needs and preferences of your students. This might involve using different communication channels for different purposes. For example, announcements could be sent via email, while quick questions could be answered through instant messaging. Consistency and clarity are paramount in effective

communication.
Remember that technology is a tool to enhance, not replace, good teaching. Focus on the pedagogical principles that underpin engaging learning, and use technology to support those principles. Don't let technology become a distraction from the core learning objectives. By thoughtfully integrating technology and focusing on creating active, collaborative, and personalized learning experiences, you can transform your e-learning environment into a dynamic and engaging space for your students.

3.1 Active Learning Strategies

Engaging students in online learning requires proactive strategies that move beyond passive content delivery. Think of it like this: you wouldn't simply lecture a group of students in a physical classroom and expect them to fully grasp complex concepts. The same principle applies to the virtual environment. Active learning methods foster deeper understanding and better retention. They demand participation, encouraging students to actively construct knowledge rather than passively receiving it. This chapter focuses on equipping you with the tools to transform your online classroom into a vibrant hub of intellectual exploration.
One highly effective technique is the flipped classroom model. Instead of delivering lectures during synchronous sessions, you provide pre-recorded video lectures or readings beforehand. This allows class time to be used for more interactive activities like

problem-solving, discussions, or project work. This approach not only caters to different learning styles but also encourages a more collaborative learning environment, fostering peer-to-peer learning and mentorship. Students prepare at their own pace, creating opportunities for deeper engagement.

Consider incorporating regular quizzes and polls throughout your online modules. These short, interactive assessments serve as powerful checkpoints, gauging student understanding and prompting immediate feedback. These aren't meant for formal grading; they are diagnostic tools. They pinpoint knowledge gaps allowing for timely intervention and adjustments to your teaching strategy. They also create a more dynamic learning experience, keeping students actively involved rather than simply passively consuming information. The immediate feedback loop helps reinforce learning and identifies areas needing further clarification.

Think beyond the typical lecture format. Structure your online courses around case studies or real-world problems. These scenarios allow students to apply theoretical knowledge to practical situations, deepening their understanding and improving critical thinking skills. Encourage students to present their findings and engage in discussions, fostering a collaborative approach to problem-solving. It's about developing a deeper understanding than just memorizing facts. It is about learning how to apply these facts in a variety of contexts and situations.

Discussions forums can be much more than just a place for students to post questions. They can become vibrant online communities. Structure discussions around thought-provoking questions or open-ended problems. Encourage debate and critical evaluation of different perspectives. Your role is to guide and facilitate, not to simply provide answers. This method promotes critical thinking, argumentation, and communication skills, all essential for academic success. Active moderation is key to ensuring respectful and productive conversations.

Another powerful tool is the use of simulations and interactive exercises. These can range from simple online games to complex virtual environments. The key is that they must be aligned with the learning objectives and cater to different learning styles. Interactive activities increase engagement and provide immediate feedback, highlighting areas for improvement. These tools can bring abstract concepts to life, transforming dry theory into engaging experiences that resonate with students.

Remember, active learning is not just about individual participation. It's about creating an environment that fosters collaboration and peer learning. Break students into small groups for online projects or discussions. Use breakout rooms during synchronous sessions. The collaborative aspect is crucial for students to learn from one another, share insights, and develop teamwork skills. This encourages students to contribute to a collective understanding of the material, further enriching the learning experience.

Finally, always solicit regular feedback from your students. Use surveys, informal polls, or even open-ended questions to understand their learning experiences. This insight helps you refine your teaching methods, ensuring your strategies remain effective and engaging. They might suggest improvements to your teaching materials or even identify areas where your approach needs refinement. Their perspective is invaluable for continuous improvement. Their feedback can directly influence how you structure future online modules and activities. The constant feedback loop is vital to successful active learning.

3.2 Collaborative Learning Techniques

Harnessing the power of collaboration in online learning environments requires a nuanced approach. Effective strategies go beyond simply assigning group projects; they involve carefully structuring interactions to maximize learning outcomes and foster a sense of community. Successful collaborative learning hinges on establishing clear goals, well-defined roles, and consistent communication channels, preventing misunderstandings and ensuring everyone contributes meaningfully.

Think of it like building a house. You can't just throw materials together and expect a sturdy structure; you need a blueprint, skilled workers, and coordinated efforts. Similarly, in online collaborative learning, a structured framework is crucial. This involves outlining project goals explicitly, defining individual roles and responsibilities, and establishing a clear timeline with specific milestones. Regular check-ins are vital – these act as quality control points, ensuring the project progresses as planned.

Effective communication is the bedrock of any successful collaboration. However, in the virtual realm, communication needs a conscious effort. Choosing the right tools is paramount. While email might suffice for announcements, platforms offering real-time interaction like discussion forums or video conferencing are often more effective for collaborative tasks. These tools allow for quick feedback, immediate clarifications, and the spontaneous exchange of ideas, mirroring the dynamic nature of in-person collaboration.

The importance of diverse perspectives cannot be overstated. Collaborative learning provides an unparalleled opportunity for students to learn from each other's strengths, filling in knowledge gaps and challenging preconceived notions. Carefully curated groups can be especially potent; mixing students with varied backgrounds, learning styles, and levels of expertise fosters a richer and more stimulating learning experience. This dynamic exchange broadens individual understanding beyond the limitations of individual study.

Assessment strategies in collaborative learning need a shift in perspective. Instead of focusing solely on individual contributions, evaluating the overall group product and the individual's role within the team is crucial. This holistic approach recognizes that success often depends on the synergy of the group. Methods like peer evaluations, self-assessments, and rubrics focusing on teamwork skills can effectively capture the full scope of student learning and contribution.

Beyond the academic, collaborative learning cultivates essential skills highly sought after in today's professional world. Effective communication, teamwork, conflict resolution, and leadership—these are all honed through the process of collaborative projects. These skills extend far beyond the immediate academic context, equipping students with valuable attributes directly applicable to future careers and personal growth. It's about fostering a holistic learning experience that extends beyond the technical aspects of the subject matter.

Careful planning and consistent facilitation are critical for the success of collaborative learning strategies. This means actively monitoring group dynamics, providing timely interventions when necessary, and offering support where needed. The instructor doesn't merely assign the task and disappear; rather, they act as a facilitator, guiding the learning process and fostering an environment of mutual support and encouragement. Think of the instructor as a coach, not just a lecturer.

Moreover, instructors must carefully consider the technologies being used. The platform needs to be intuitive, user-friendly, and have the capabilities required for effective collaboration. This includes features for file sharing, version control, communication tools, and mechanisms for providing feedback. Poor technology choices can quickly derail even the most well-designed collaborative activity.

Finally, remember that effective collaborative learning isn't about simply throwing students into groups and hoping for the best. It requires careful consideration of pedagogical principles, strategic planning, active facilitation, and a mindful approach to assessment. By adopting these strategies, educators can unlock the immense potential of collaborative learning, transforming the online learning experience into a dynamic, engaging, and richly rewarding endeavor for all involved. The rewards extend beyond improved academic performance; they cultivate valuable life skills, fostering a sense of community and preparing students for success in a collaborative world.

3.3 Gamification in E-Learning

Gamification leverages game-design elements in non-game contexts, offering a powerful tool to boost engagement and motivation in e-learning. Think points, badges, leaderboards – familiar elements that tap into our innate desire for accomplishment and social recognition. By strategically integrating these mechanics, educators can transform mundane tasks into rewarding experiences, fostering a more positive and productive learning environment. Effective gamification isn't about turning the entire course into a video game; it's about carefully selecting appropriate elements to enhance specific learning objectives.

Consider the potential impact on student motivation. A simple points system, awarding points for completing assignments or participating in discussions, can provide a tangible sense of progress, keeping students motivated and on track. Badges, representing mastery of specific skills or concepts, offer visual reinforcement of achievement and a sense of accomplishment, subtly encouraging continued learning. Leaderboards, while potentially controversial due to the competitive element, can spark healthy competition and encourage peer learning when implemented thoughtfully, focusing on collaboration rather than pure competition.

The key to successful gamification lies in its alignment with learning objectives. Avoid arbitrary rewards; every game element should directly support the course's goals. For instance, if the course objective is to improve critical thinking skills, the rewards system might focus on rewarding well-reasoned arguments and insightful contributions to discussions rather than simply completion of tasks. This ensures that the game mechanics reinforce the learning process, rather than distracting from it. Remember, the goal isn't just to keep students engaged, but to improve their learning outcomes.

The design of the gamified elements themselves is crucial. Overly complex systems can overwhelm students and detract from the learning experience. Simplicity and clarity are paramount; the mechanics should be intuitive and easy to understand. Consider the visual design as well: appealing graphics, animations, and interactive elements can enhance engagement. A visually stimulating learning environment can significantly impact student motivation and learning effectiveness. Make sure your visuals are accessible to all students, keeping in mind visual impairments and color blindness.

Beyond points and badges, consider incorporating storytelling elements into the learning experience. Frame the course content as a quest or a challenge, where students must overcome obstacles and solve puzzles to achieve their goals. This narrative structure can increase engagement and provide a more immersive learning experience. Students may also be intrinsically motivated to solve these engaging puzzles, making even complex material seem more approachable. Make sure your narratives are inclusive and avoid harmful stereotypes.

Integrating social elements is another crucial step. Peer-to-peer interaction, whether through collaborative projects, group discussions, or shared leaderboards, can foster a sense of community and belonging within the virtual classroom. This social interaction can be incredibly motivating, especially for students who benefit from a sense of connection and support. Careful moderation is necessary to avoid toxic competitiveness or negative interactions. Set clear guidelines and expectations for online conduct.

Finally, remember that gamification is just one piece of the puzzle. It's most effective when integrated into a broader strategy for creating engaging and effective online learning experiences. Combine it with other strategies discussed in previous chapters such as active learning techniques, collaborative learning, and personalized learning paths for a well-rounded and effective approach. Continuously assess the effectiveness of your gamification strategy; monitor student engagement and feedback to refine your approach. Regularly review your metrics and make adjustments as needed. What works for one group might not work for another.

Remember, the ultimate goal isn't to create the most entertaining online course; it's to create the most effective one. Gamification can be a valuable tool in achieving this goal, but it's only one component of a larger strategy. By carefully considering the learning objectives, choosing the right game mechanics, and incorporating social and narrative elements, educators can create engaging and rewarding online learning experiences that promote student success. It's a powerful tool, but its effectiveness relies heavily on thoughtful planning and implementation. Don't be afraid to experiment and adapt your approach based on your students' feedback and your own observations of their engagement.

3.4 Assessment & Feedback Methods

Effective assessment is the cornerstone of successful e-learning. It's not simply about grading; it's about providing valuable feedback that guides student learning and improves instructional design. Choosing the right assessment methods is crucial; a mismatch between assessment and learning objectives can lead to frustration and inaccurate evaluation of student understanding. Therefore, carefully consider the learning outcomes you hope to achieve before settling on your assessment strategy. Multiple-choice questions are easy to grade but rarely assess critical thinking skills. Essays, on the other hand, allow for more in-depth analysis but are more time-consuming to evaluate.

A balanced approach, utilizing a variety of assessment methods, often yields the best results, offering a more comprehensive view of student comprehension.

To truly gauge student understanding, incorporate various assessment formats. Short quizzes, for instance, test immediate recall and knowledge application, allowing for quick feedback. Longer assignments, such as projects or research papers, evaluate students' ability to synthesize information, conduct research, and communicate their findings effectively. These assignments also provide opportunities for more extensive feedback, addressing individual student challenges more thoroughly. Remember that a diverse assessment strategy should include both formative and summative assessments. Formative assessments, like in-class activities or peer reviews, provide ongoing feedback, allowing for adjustments to instruction and student learning strategies. Summative assessments, such as final exams or major projects, evaluate overall learning outcomes at the end of a course or module.

Providing feedback is equally crucial. Simply stating a grade is insufficient. Constructive feedback offers specific suggestions for improvement, highlighting both strengths and weaknesses. Specific, actionable feedback is more effective than general comments. For instance, instead of saying "Good job," consider saying, "Your analysis of the data was thorough, but you could strengthen your conclusion by incorporating evidence X." The goal isn't to discourage students but to equip them with the tools to succeed. Timely feedback is also essential. Delayed feedback loses its impact, making it less likely that students will apply the suggestions provided. Aim for timely feedback; the sooner students receive feedback, the more likely they are to integrate it into their future work.

Technology can significantly enhance the assessment and feedback process. Learning management systems (LMS) offer automated grading for multiple-choice questions and other objective assessments. This frees up instructors' time, allowing for more focused feedback on subjective assignments. Furthermore, LMS platforms often facilitate communication between instructors and students, allowing for easier feedback dissemination and clarifying questions. These systems often include features that allow for individualized feedback using rubrics and annotations. This personalized approach ensures students receive tailored guidance specific to their performance and needs. However, remember that technology should enhance, not replace, human interaction. Personalization via technology offers efficiency, but direct feedback from the instructor provides the most impactful learning experience.

Consider the use of peer assessment. Students reviewing their classmates' work fosters critical thinking skills and promotes active learning. The process also provides students with valuable perspectives on their own work and offers an opportunity to receive constructive criticism from their peers. However, carefully design the peer assessment process to ensure fairness and accuracy, providing clear guidelines and criteria for evaluation. This avoids misunderstandings and reduces the likelihood of biased or unfair reviews. Implementing a well-structured peer assessment system enhances learning for both the assessor and the assessed, fostering a collaborative learning environment while reducing the instructor's workload. You can then use this feedback to gain insights into individual student learning needs, allowing for adjustments in instructional strategies.

Ultimately, effective assessment and feedback methods are integral to a successful e-learning experience. They are not simply a means of measuring student achievement but rather a crucial tool for fostering continuous improvement in both teaching and learning. By carefully selecting assessments, providing timely and constructive feedback, and leveraging technology effectively, you create a dynamic and supportive learning environment. Remembering to focus on student learning and growth rather than solely on grades creates a more positive and engaging learning experience. It is through continuous evaluation and improvement of assessment strategies that you optimize the effectiveness of your e-learning program. This continual refinement of your techniques ensures the enduring legacy of impactful e-learning in higher education.

3.5 Personalized Learning Paths

Tailoring the educational journey to each student's unique needs is paramount in maximizing e-learning's potential. Personalized learning paths, far from being a mere buzzword, represent a fundamental shift in how we approach online education, moving away from a one-size-fits-all model to a more nuanced, adaptive approach. This necessitates a detailed understanding of individual learning styles, preferences, and prior knowledge. Effective implementation involves leveraging technology to create flexible pathways that accommodate diverse learning speeds and accommodate varying levels of prior experience.

This personalized approach requires careful planning. Initial assessments are crucial; they shouldn't simply be tests, but rather comprehensive evaluations that gauge not just knowledge but also learning styles, strengths, weaknesses, and overall learning preferences. Consider using a variety of assessment methods, such as pre-tests, questionnaires on learning habits, and self-reflection exercises. Analyze the data gathered to create a detailed profile for each learner. This profile will serve as the foundation for the tailored learning journey. Remember, the goal is to understand how each student learns most effectively, not just what they already know.

The creation of personalized learning paths isn't a one-off task, but an ongoing process of adaptation and refinement. Regular checks are critical. Track student progress through integrated learning management systems (LMS). These systems offer valuable insights into student engagement, the time spent on specific modules, and the types of content the student interacts with most. This data-driven approach helps identify areas where the student excels, as well as those that require additional support or a modified approach. Don't hesitate to adjust the path; flexibility is key. Perhaps a student excels with videos, but struggles with text-heavy modules. Adapt the pathway to emphasize visual resources where necessary.

Technology plays a crucial role in facilitating these personalized paths. Adaptive learning platforms, for instance, automatically adjust the difficulty and content based on the student's progress and performance. They provide immediate feedback, identify knowledge gaps, and suggest additional resources to address those gaps. These tools not only personalize the learning experience, but also free up the instructor's time, allowing them to focus on providing individualized support and mentoring rather than creating repetitive tasks. Effective use of technology means integrating it seamlessly into the curriculum, ensuring it enhances, not hinders, the learning process. It's not just about using technology for technology's sake; it's about strategically integrating it to support personalized learning.

Remember that personalization extends beyond simply altering the difficulty level. It encompasses the choice of learning materials, the style of instruction, and even the timing and pacing of the course. Some students might thrive in a structured environment with clearly defined goals and timelines, while others may benefit from a more flexible, self-directed approach. Offering options is critical. Providing varied materials, such as interactive simulations, podcasts, video lectures, and traditional text-based readings, allows students to engage with the information in ways that resonate most with their individual learning styles. This variety also caters to students with different preferences and needs.

It is crucial to consider the human element in this technological approach. Personalized learning is not solely about algorithms and adaptive platforms; it requires consistent, supportive interaction with the instructor. Regular check-ins, personalized feedback, and opportunities for one-on-one guidance are crucial in fostering a positive and productive learning environment. The instructor acts as a mentor, guiding students through their individual learning journeys, providing encouragement and support when needed. A sense of community is also important; create spaces for students to interact with each other and share their experiences. The goal is to create a supportive and inclusive online learning community. Remember that technology should enhance the human connection, not replace it.

Finally, effective implementation requires ongoing evaluation and refinement. Gather student feedback regularly through surveys, focus groups, and informal conversations. Analyze the data collected to identify areas for improvement. Refining the personalized learning paths is an iterative process; it is about continuous improvement to ensure that the system remains responsive to the evolving needs and preferences of the students. This requires a commitment to continuous improvement and a willingness to adapt and evolve the approach over time. Regularly reviewing and refining the process is a necessity to fully reap the benefits of personalized learning paths.

The data you collect will offer invaluable insights into the effectiveness of your approach and reveal areas requiring adjustment for an even more optimized learning experience. This ongoing assessment helps ensure that the personalized learning paths effectively meet the needs of each individual student.

4 Best Practices for Online Instruction

Clear, concise communication is paramount. Avoid overly technical jargon; instead, prioritize simple, direct language that all students can readily understand. Regularly check for comprehension—ask questions, solicit feedback, and actively monitor student participation to gauge understanding and identify any confusion early. Remember, your online persona is just as important as your in-person presence. Maintain a professional, approachable demeanor; encourage questions and create a welcoming atmosphere to foster trust and open communication. This promotes a sense of community and helps students feel comfortable engaging in discussions.

Regular and consistent communication is crucial. Schedule regular office hours, respond promptly to emails and forum posts, and provide timely feedback on assignments. Establishing a clear communication plan at the beginning of the course, outlining expectations and methods, sets the stage for positive interaction and prevents misunderstandings. Consider using multiple communication channels—email for formal announcements, discussion forums for collaborative learning, and instant messaging for quick queries. The key is consistent accessibility and a rapid response to student needs. Proactive student support is more than just responding to problems; it's about anticipating them and preventing them entirely. Clearly outline your expectations, policies, and course procedures early on, minimizing uncertainty and potential issues. Provide readily accessible resources, such as FAQs, tutorials, and technical support information, so students can troubleshoot common problems independently. Consider establishing a dedicated online help desk or forum for technical or course-related assistance. Regularly monitor student progress through assessments and participation levels, intervening early if you detect signs of disengagement or difficulty.

Engaging students in the online environment requires thoughtful planning and intentional strategies. Incorporate interactive elements like polls, quizzes, and collaborative projects to actively involve students in the learning process. Break up large content chunks into smaller, more manageable modules to improve focus and reduce cognitive overload. Consider using multimedia resources—videos, images, and audio—to diversify learning materials and cater to various learning styles. Foster a sense of community by creating opportunities for peer-to-peer interaction; facilitate group discussions, encourage collaborative projects, and promote the use of online forums.

Create opportunities for active learning. Instead of simply lecturing, design activities that require students to apply what they've learned. Consider incorporating case studies, simulations, problem-solving exercises, and group discussions to enhance engagement and retention. Regular feedback is vital. Provide constructive criticism on assignments and assessments, focusing on both strengths and weaknesses. Promptly address any misconceptions or gaps in understanding. This timely feedback cycle allows students to correct mistakes and improve their learning. Offer personalized feedback whenever possible; tailoring comments to individual student needs demonstrates your dedication and enhances student learning.

Remember that technology is a tool, not the focus. Choose technology that supports your pedagogical goals, not the other way around. Prioritize user-friendliness and accessibility. Avoid overwhelming students with too many platforms or complex technologies. Instead, select a few well-integrated tools that effectively support your teaching and learning goals. Always ensure that the technology supports inclusivity—for example, ensure that all materials are accessible to students with disabilities, and provide appropriate alternative formats if necessary. Regular maintenance and updates of the chosen technology are important for uninterrupted usage.

Building a supportive online learning community is key to student success. Facilitate opportunities for interaction and collaboration among students. Encourage peer-to-peer learning, and create a safe space where students feel comfortable asking questions and sharing their perspectives. Regularly check in with students to gauge their engagement and address any concerns. Make yourself accessible and responsive to their needs. Respond promptly to questions, provide timely feedback, and cultivate an atmosphere where students feel valued and respected. A strong sense of community helps to combat feelings of isolation, so common in the online learning experience. Promoting a positive learning environment increases student engagement and significantly improves overall learning outcomes.

Finally, continuously evaluate and refine your teaching practices. Seek feedback from students on what works well and what could be improved. Analyze assessment data to track student progress and identify areas where you can provide more support. Stay current with best practices in online instruction by engaging in professional development opportunities and attending conferences. Remember that teaching in the online environment is an iterative process; constantly adapting and refining your approaches based on evidence and feedback is vital for providing high-quality online instruction. This ensures that your online courses are continuously evolving and improving, meeting the evolving needs of both students and educators.

4.1 Effective Communication

Clear, concise communication is paramount in online education. Ambiguity leads to confusion, hindering learning. Students need precise instructions, readily available resources, and prompt feedback to thrive in a virtual environment. Without it, frustration mounts, participation wanes, and ultimately, learning suffers. Therefore, prioritize clarity above all else. Use simple language, avoiding jargon or overly technical terms that could leave students feeling lost and discouraged.

Regular and consistent communication is key. Don't just broadcast information; engage in a two-way dialogue. This involves actively seeking student input, addressing their questions promptly, and fostering a sense of community. Consider using discussion forums, interactive polls, or even virtual office hours to encourage interaction and build rapport. Respond to emails and messages within a reasonable timeframe, demonstrating your commitment to their success. Remember, consistent communication builds trust and reinforces your role as a facilitator of learning.

Effective communication extends beyond simple information exchange. It involves actively listening to students' concerns and understanding their perspectives. Create opportunities for them to express themselves, whether through assignments, discussions, or feedback forms. Pay attention to the nuances in their communication, recognizing that sometimes, what is left unsaid can be equally important. Showing empathy and understanding fosters a supportive learning environment where students feel comfortable seeking help and expressing their opinions. This leads to better engagement and improved learning outcomes.

Visual communication should not be overlooked. In the online realm, visual aids can significantly enhance understanding. Use high-quality images, videos, and interactive presentations to complement your textual content. Remember, visuals cater to different learning styles and make complex information more easily digestible. A well-designed infographic, for example, can convey information more effectively than a dense paragraph of text. Consider this when preparing your course materials. Think about the specific needs of your online students; are they visual learners? Do they respond better to a particular style of visual aid?

Beyond the content itself, the manner of communication is crucial. Your tone should be approachable and encouraging. Avoid overly formal or authoritative language, which can create a distance between you and your students. Strive for a conversational style that promotes a sense of inclusivity and mutual respect. This involves using inclusive language, avoiding jargon, and showing genuine interest in your students' learning journey. This is especially vital in a virtual environment where nonverbal cues might be less readily available.

Think about the accessibility of your communication. Ensure all materials are easily accessible to students with varying levels of technological proficiency. Provide clear instructions on how to access and use different platforms and tools. Furthermore, utilize multiple channels of communication, recognizing that some students may prefer email, while others might engage more readily through discussion boards or instant messaging. Providing multiple avenues allows students to select the communication method most suited to their preferences and needs. This contributes to a more inclusive learning environment where everyone can participate effectively. Furthermore, carefully consider the timing of your communication. Avoid bombarding students with information all at once. Space out announcements, assignments, and feedback strategically to prevent information overload. Use announcements effectively; don't use them for minor details but focus on key updates or reminders. Be mindful of students' time zones, especially in asynchronous learning environments. Try to avoid scheduling live sessions at times that are inconvenient for a significant portion of your students. These are small, yet important, considerations in building a functional and supportive learning experience.

Finally, remember that effective communication is a continuous process. Regularly solicit student feedback to assess the effectiveness of your communication strategies. Ask for their input on the clarity of instructions, the accessibility of materials, and the overall learning experience. Use this information to refine your approach, ensuring that your communication is always clear, concise, and supportive. This feedback loop is essential to improve the learning experience and create a more positive and engaging online environment for all involved. Continual adaptation and improvement will significantly benefit both your teaching and your students' learning outcomes.

4.2 Student Support & Engagement

Proactive student support is paramount; it's not merely reactive troubleshooting. Think of it as preventative maintenance for your virtual classroom. Regularly check-in with students, not just for grades, but for their overall well-being and comprehension. A simple, weekly email expressing genuine interest goes a long way. Consider incorporating brief, informal surveys to gauge student understanding and pinpoint areas needing clarification. This shows you're invested in their success beyond the assignment grades. This fosters a sense of community. Students who feel heard are more engaged. They're more likely to participate actively and contribute meaningfully. That creates a richer learning experience for everyone.

Establishing clear communication channels is essential. Regular office hours, accessible via video conferencing, offer a convenient way for students to receive personalized guidance. Provide prompt and constructive feedback on assignments. Don't just offer a grade; provide specific suggestions for improvement. This demonstrates your commitment to their academic growth and provides valuable learning opportunities. This helps students improve their weaknesses and celebrate their strengths. Remember, timely feedback is more effective than delayed feedback. If there are extenuating circumstances, maintain consistent open communication. Understanding circumstances can reduce stress for students while also making it easier to develop alternative solutions.

Creating a supportive online community goes beyond simple forums. Organize virtual study groups or collaborative projects. These activities promote peer learning and interaction, which helps ease feelings of isolation often associated with online learning. Facilitate discussions through engaging prompts that encourage students to share their perspectives and build upon one another's ideas. Encourage participation, but avoid making participation mandatory. Forcing participation can often lead to resentment instead of collaboration. Encourage students to help one another; facilitate that peer learning. Promote a culture of mutual respect and support. This creates a more inclusive environment.

Regularly assess student engagement. Monitor participation in discussions, assignment submissions, and attendance in virtual sessions. Look for patterns or trends that may indicate students who are struggling or disengaging. Address these issues proactively, and make sure students know they are not alone, and you will assist them. This is not just about identifying problems; it's about offering effective solutions tailored to individual student needs. Remember that engagement is not just about activity, but about meaningful interaction.

Consider utilizing technology to enhance student support. Learning management systems often offer tools for tracking student progress, identifying at-risk students, and facilitating communication. Many also feature built-in support tools, such as tutorials, FAQs, and helpdesk support. Integrate these systems strategically into your teaching methodology. These tools can streamline administrative tasks, allowing you more time for individual student interaction. Students will benefit from your time investment, which can mean the difference between success and struggle. The technology should supplement, not replace, your role as an instructor.

Remember, student engagement isn't a passive state; it's an active process that requires your consistent effort. Regularly review and adapt your support strategies based on student feedback and your observations. This iterative approach ensures you are continually improving the overall learning environment. Creating an inclusive and supportive online learning environment is crucial for student success. By fostering a sense of community, providing personalized support, and utilizing technology effectively, you can cultivate an environment where every student feels valued, engaged, and empowered. The key is consistent and attentive support. It's a commitment that pays dividends in both student achievement and your own satisfaction as an instructor. Remember, your role extends beyond dispensing information; it's about nurturing learners and helping them thrive. Your investment in student support ultimately enhances the effectiveness of the entire e-learning experience. The most effective e-learning is built upon a foundation of strong student support and genuine engagement.

5 Accessibility & Inclusivity in E-Learning

Designing accessible and inclusive e-learning environments is paramount. It's not merely about compliance; it's about fostering a learning community where every student feels valued, respected, and empowered to reach their full potential. This requires a proactive and multifaceted approach, going beyond simply meeting minimum accessibility standards.

Consider the diverse needs of your student population. This goes beyond physical disabilities; it encompasses learning differences like dyslexia or ADHD, neurodiversity, varying levels of digital literacy, and socio-economic backgrounds affecting access to technology and reliable internet. Understanding these varied needs forms the bedrock of truly inclusive design.

Universal Design for Learning (UDL) provides a powerful framework. UDL emphasizes providing multiple means of representation (presenting information in diverse formats), action and expression (allowing students to demonstrate their learning in varied ways), and engagement (motivating and challenging learners through varied approaches). Implementing UDL principles ensures materials are adaptable to diverse learning styles and needs.

Assistive technologies play a vital role. This includes screen readers for visually impaired students, text-to-speech software for those with reading difficulties, and keyboard navigation for those with motor impairments. It's crucial to ensure your learning materials and platforms are compatible with a broad range of assistive technologies. Thorough testing with these technologies is critical before launching your course.

Creating inclusive online communities is essential. Foster a culture of respect and understanding where students feel comfortable sharing their needs and seeking support. Facilitate communication through various methods – text, audio, video – to accommodate different communication preferences. Encourage peer-to-peer support and collaboration to promote a sense of belonging. Clearly defined communication guidelines and expectations can help build a more inclusive environment.

Promoting accessibility goes beyond the technical aspects. It's about fostering a culture of inclusivity where accessibility is considered throughout the entire e-learning design process, from content creation to assessment. Train instructors on inclusive teaching practices and provide resources to support their efforts. Regularly solicit feedback from students with disabilities to identify areas for improvement and ensure your e-learning environment is genuinely

accessible and welcoming to all.
Remember that accessibility is an ongoing process, not a one-time fix. Regularly review and update your e-learning materials and practices to reflect evolving accessibility standards and best practices. Keep abreast of emerging assistive technologies and incorporate them into your design. Remember that the ultimate goal is to create a rich and equitable learning experience for all students, regardless of their abilities or backgrounds. This will ultimately lead to a more engaging and enriching learning journey for everyone involved. Accessibility is not just a box to tick; it is a commitment to inclusivity, equity, and equal opportunity in education. It's about empowering all learners to participate fully and achieve their academic goals. Prioritize it. It's the right thing to do, and it makes for a better learning experience for everyone. Accessibility should be an integral part of the design process, not an afterthought, ensuring a consistent and equitable experience for all students.

5.1 Universal Design for Learning

Universal Design for Learning (UDL) is a framework that guides the creation of inclusive learning environments. It's not merely about accommodating students with disabilities; it's about designing learning experiences that are accessible and engaging for all learners, regardless of their background, learning style, or abilities. Think of it as proactively building flexibility and choice into your online courses, ensuring everyone can participate and succeed.

The core of UDL rests on three main principles: multiple means of representation, multiple means of action and expression, and multiple means of engagement. Multiple means of representation address how information is presented to learners. Instead of relying solely on text, consider incorporating videos, audio recordings, interactive simulations, and visual aids. Offer diverse formats to cater to various learning preferences and needs. For example, providing transcripts for videos, offering text-to-speech options, and using clear, concise language are all part of this. Remember, visual learners might benefit from diagrams and charts, while auditory learners might prefer podcasts or discussions.

Next, multiple means of action and expression focus on how students demonstrate their learning. Don't limit assessment to traditional exams. Instead, provide a range of options, such as presentations, essays, projects, portfolios, and even digital storytelling. Allow students to showcase their understanding in ways that best suit their strengths. Some students might excel at writing, others might be more adept at visual communication, and others may flourish through practical application. Offering diverse assessment methods caters to these varied strengths. Think about allowing students to choose the format for their assignments, or providing different levels of scaffolding to support different learners.

Finally, multiple means of engagement address why learners are motivated to participate. Tap into students' intrinsic motivation by offering choices, fostering collaboration, and creating a sense of community. Provide opportunities for autonomy and self-regulation, allowing learners to control the pace and style of their learning. Consider incorporating real-world applications, relevant case studies, and opportunities for feedback to maintain their engagement. This could mean allowing students to work on projects that matter to them, or offering flexible deadlines.

Implementing UDL in your online courses requires careful planning and a shift in mindset. It's not a one-size-fits-all solution, and it requires careful consideration of the specific needs of your student population. Start by analyzing your existing course materials and identifying areas where you could increase accessibility. Consider how the materials are presented, how students are assessed and how the course engages them. This may involve replacing lengthy, dense text with shorter, more concise paragraphs, using clear and consistent headings and subheadings and adding more images, videos, or interactive elements.

Think creatively about the kinds of assessments you use. Could you replace a traditional essay with a presentation, a podcast, or a project-based assignment? Would allowing students to choose their preferred method for demonstrating their understanding lead to improved participation and performance? Consider the overall accessibility of your online learning environment. Are your materials compatible with assistive technologies? Is your course accessible to students with visual or hearing impairments?

Remember, UDL is an iterative process. You will likely need to refine and improve your approach over time as you learn more about the learning styles and preferences of your students. Observe the different ways they learn and adapt, based on your observation, adjust the teaching and materials for them. This proactive approach to inclusive design not only benefits students with disabilities but also creates a more enriching and engaging learning experience for all learners. Regular feedback is crucial, asking for student input on accessibility and usability through surveys or informal feedback sessions provides invaluable insights.

By adopting UDL principles, you can create online learning environments that are truly inclusive, engaging, and effective for all students. This means you are not only building a better online experience, but you're also building better futures for your learners, equipping them with the skills they need to succeed in a diverse and complex world. This shift in perspective—from accommodating to proactively designing for accessibility—marks a significant advancement in creating a truly inclusive educational setting for every individual learner in your virtual classroom. Remember that this commitment is not merely a compliance issue but a genuine commitment to fostering equity and maximizing every student's potential. Continuous evaluation of teaching practices and providing resources that are tailored to the specific needs of every student demonstrates your support for their success.

5.2 Addressing Diverse Needs

Catering to diverse learners requires a multifaceted approach. We must move beyond the simplistic notion of "one size fits all" and embrace the reality of varied learning styles, technological access, and individual needs. This necessitates a pedagogical shift, one that prioritizes inclusivity and accessibility as core design principles, rather than afterthoughts.

Consider, for example, the student with a visual impairment. Simply providing text-based materials is insufficient. We need to integrate alternative formats like audio descriptions, screen reader compatibility, and transcripts of video lectures. These aren't mere additions; they are fundamental components of a truly accessible learning experience. Failing to provide these options actively excludes a significant population, negating the very inclusivity we strive for.

Furthermore, cognitive differences demand thoughtful consideration. Students with ADHD might benefit from shorter, more focused learning modules, interspersed with frequent breaks and interactive elements to maintain engagement. Similarly, students with dyslexia might require dyslexia-friendly fonts, extended time for assessments, and alternative methods of demonstrating their understanding. The key is to provide flexible options that cater to their individual learning preferences and challenges without compromising academic rigor.

Beyond individual needs, we must also consider the broader spectrum of technological access. Not all students have reliable high-speed internet or access to sophisticated devices. This disparity necessitates the creation of materials that function effectively on a range of devices and bandwidths. This might involve creating smaller file sizes for downloads, utilizing accessible web design principles, and offering alternative methods of accessing materials, such as downloadable PDFs or offline access options. It's not just about providing digital content; it's about ensuring equitable access to that content.

Effective communication is paramount. Clear and concise instructions, multiple methods of contact with instructors, and open communication channels are essential for all students but even more critical for those who might require extra support or clarification. This extends to ensuring materials are written in clear, simple language, avoiding jargon and unnecessary complexity. Regular check-ins, both individually and with the class as a whole, help identify potential challenges early on, enabling timely interventions. We should not underestimate the power of proactive communication in building a supportive learning environment.

The design of assessments should also reflect this commitment to diversity. Offering varied assessment methods—such as essays, presentations, projects, and multiple-choice questions—allows students to showcase their understanding in ways that best suit their strengths and abilities. Furthermore, providing clear rubrics and feedback mechanisms helps students understand the expectations and identify areas for improvement, fostering a growth mindset. In addition, implementing flexible scheduling and deadlines demonstrates empathy and understanding of diverse circumstances and responsibilities that students may face outside of their academic life.

Incorporating universal design for learning (UDL) principles serves as a valuable framework. UDL provides a structured approach to developing curriculum and learning materials that are inherently accessible to all learners. By providing multiple means of representation, action and expression, and engagement, we create a learning environment where every student can thrive, regardless of their individual needs or background. It's about anticipating and proactively addressing the diverse needs of our learners before encountering challenges.

Building a truly inclusive online learning environment requires a collective commitment. This involves faculty development, institutional support, and ongoing evaluation to ensure we are meeting the needs of all students. It's an iterative process of continuous improvement, guided by student feedback and a commitment to equity. The goal is not just to provide access but to create an environment where all students feel valued, supported, and empowered to reach their full potential. The journey towards inclusive e-learning isn't a destination; it's an ongoing process that requires continuous reflection and adaptation. Ultimately, the success of our online programs hinges on our ability to create truly inclusive learning communities where every student has the opportunity to thrive.

5.3 Assistive Technologies

Assistive technologies are crucial for creating truly inclusive online learning environments. They empower students with disabilities to fully participate and succeed in higher education. This isn't simply about compliance; it's about unlocking potential and fostering a sense of belonging for all learners. A commitment to accessibility is a commitment to educational equity.

Consider screen readers, for example. These invaluable tools convert text to speech, allowing visually impaired students to access course materials, participate in discussions, and complete assignments. Properly formatted documents, adhering to accessibility standards like WCAG, are essential for effective screen reader use. Think of the frustration a student might experience trying to navigate poorly structured PDFs—a frustration that can be easily avoided with thoughtful design. Furthermore, training faculty on the effective use and integration of screen readers, along with best practices for creating accessible digital content, is equally important to ensure success. This isn't just about providing the technology; it's about building a supportive ecosystem around it.

Beyond screen readers, text-to-speech software offers broader benefits. Students with dyslexia or other learning differences can use this technology to hear materials read aloud, enhancing comprehension and reducing cognitive load. Similarly, speech-to-text software allows students with physical limitations to contribute effectively to class discussions and complete written assignments. Imagine the empowerment a student feels when they can effortlessly dictate their thoughts, transforming spoken words into typed text. This is more than convenience; it's about equal opportunity. This is crucial for students who may struggle with traditional input methods.

Captioning and transcription services are equally transformative, opening up multimedia content to students who are deaf or hard of hearing. This isn't just about adding subtitles; it's about creating an immersive experience where all students can follow along. Accurate and timely captions and transcripts are vital—inaccurate or delayed ones can hinder understanding and participation. Moreover, providing captioned videos and transcripts offers significant benefits to all students, including those who may be learning English as a second language or who prefer to review materials later at their own pace. The inclusion of captions ensures information is accessible to a larger audience. Consider the investment in reliable transcription services as an investment in quality education.

Alternative input devices, like eye-tracking technology and specialized keyboards, can assist students with motor impairments. These allow learners to interact with technology in a way that best suits their needs. It's a matter of enabling the learning process through technological innovation. The integration of such technologies might require some initial investment and training but offers substantial long-term benefits. This is an opportunity to showcase the institution's commitment to inclusivity. The benefits extend to everyone, fostering a more supportive learning community. The return on investment in terms of student success and satisfaction makes it worthwhile.

Remember, assistive technologies are not one-size-fits-all solutions. Individual needs vary widely. Therefore, a multi-pronged approach to accessibility is crucial, offering a diverse range of tools and support systems to meet various needs. This includes personalized consultations with assistive technology specialists, dedicated training for both faculty and students, and ongoing technical support. It's about creating a system that adapts to the learner, not the other way around. Effective assistive technology implementation requires a deep understanding of individual needs and a commitment to continuous improvement.

Furthermore, it's not enough to simply provide the technology. Faculty need training on how to effectively integrate these tools into their teaching practices. This ensures that technology enhances, not hinders, the learning process. A robust training program can be vital for successful implementation of assistive technologies. It ensures that the technology is used appropriately and effectively. The faculty's role in this process shouldn't be overlooked.

Creating a truly accessible online learning environment is not a destination, but a continuous process of evaluation and improvement. Regularly soliciting feedback from students with disabilities is essential to identify areas for enhancement and ensure that the assistive technologies provided are both relevant and effective. This ongoing assessment allows the institution to continuously adapt and improve its accessibility measures, meeting the changing needs of the student body. The ongoing feedback loop ensures continuous improvement. Regular evaluations are essential.

5.4 Creating Inclusive Online Communities

Building a truly inclusive online community requires careful planning and consistent effort. It's not simply about adding accessibility features; it's about fostering a sense of belonging where every participant feels valued and respected, regardless of their background, abilities, or learning style. Consider your community's structure; how easily can individuals participate and contribute? Think about the ease of navigation and information access. A well-designed interface is critical for inclusivity.

Clear and concise communication is paramount. Use plain language and avoid jargon. Provide transcripts for videos and alternative text for images. Remember that diverse learners might process information in different ways. Multiple formats and presentation styles can ensure everyone can access and understand the material presented. For example, offer both text and audio versions of announcements. Consider using different fonts and sizes to accommodate visual differences.

Active moderation is essential. Establish clear community guidelines that promote respect and inclusivity. These guidelines should address harassment, discrimination, and other forms of inappropriate behavior. Enforce these rules consistently and fairly to maintain a safe and welcoming environment. Regularly monitor discussions to prevent negativity or harmful exchanges from escalating. Respond promptly to concerns and report any violations that may occur. This proactive approach will strengthen the community.

Encourage participation from diverse members. Actively recruit individuals from different backgrounds and experiences. Provide opportunities for everyone to share their perspectives and contribute to discussions. Create dedicated spaces for smaller group discussions, fostering a sense of camaraderie among members who may share similar backgrounds or interests. Promote these channels and actively support their development to reach a greater segment of the community.

Use technology to enhance inclusivity. Explore assistive technologies such as screen readers, text-to-speech software, and captioning tools. Ensure all your resources and learning materials are accessible to individuals with disabilities. For example, using closed captions will make videos accessible to deaf and hard-of-hearing learners. This inclusive approach enhances participation for all users.

Seek feedback regularly. Collect feedback from your community members to identify areas for improvement. Conduct surveys, polls, and focus groups to understand their experiences and assess the inclusiveness of your community. Analyze the data to create more effective strategies for fostering inclusion. Be open to suggestions and modifications; flexibility and adaptability are essential components of this process.

Remember that inclusivity is an ongoing process. It requires consistent effort and a commitment to creating a welcoming environment. Regularly review and update your policies and practices to ensure they continue to meet the needs of your diverse community. By prioritizing inclusivity, you will create a richer, more vibrant, and valuable learning experience for everyone. This proactive approach to building community strengthens the overall atmosphere and improves the participation level for all users. This fosters a higher sense of belonging and increases the likelihood that users will continue to participate in online discussions. This dedication is reflected in the positive reception by the users and leads to a collaborative and productive online community.

Continuously evaluate the effectiveness of your inclusivity initiatives. Track metrics such as participation rates, feedback scores, and overall community sentiment. Use data to inform decision-making and adjust strategies as needed. Remember that inclusion is not a destination but a journey. The continuous process of adaptation and improvement will shape and refine the effectiveness of the community, fostering a strong and supportive environment. By actively pursuing this ongoing evaluation process, the community will remain vibrant and engaging. This commitment to growth and adaptation will ensure its success in the long term and promote continuous learning. Regular reassessment will guarantee that the community always meets the needs of its members.

Finally, remember that building an inclusive online community is a collaborative effort. Involve community members in the process of shaping and defining inclusivity. By working together, you can create a learning environment where everyone feels welcome, respected, and valued. The shared ownership of the community fosters a feeling of responsibility and encourages participation from all segments of the membership, leading to a stronger and more dynamic community. Active involvement and collaborative effort is vital in building a thriving community. This shared vision and engagement will lead to increased participation from all members and a stronger sense of belonging.

5.5 Promoting Accessibility

Creating truly accessible e-learning requires a multifaceted approach. It's not just about adding captions to videos; it's about fundamentally reshaping how we design and deliver online courses. Consider the diverse needs of your learners – students with visual, auditory, motor, cognitive, and learning disabilities, as well as those from diverse linguistic and cultural backgrounds. Each presents unique challenges and necessitates thoughtful solutions. For example, a student with dyslexia might benefit from text-to-speech software, while a student with a visual impairment might require screen reader compatibility. Failing to address these needs creates a significant barrier to participation and limits the potential of online learning.

Designing for accessibility isn't an afterthought; it's integral to the design process. Universal Design for Learning (UDL) provides a valuable framework. UDL emphasizes providing multiple means of representation, action and expression, and engagement. This means offering information in various formats—text, audio, video, images—allowing students to demonstrate their learning in different ways, and fostering motivation and engagement through varied activities. Implementing UDL ensures the course is inherently accessible, eliminating the need for extensive modifications later. Start by creating clear, concise content, using headings, subheadings, and bullet points to improve readability. Ensure adequate color contrast between text and background for those with visual impairments.

Assistive technologies play a crucial role in bridging the accessibility gap. These technologies, including screen readers, text-to-speech software, and alternative input devices, empower students with disabilities to access and interact with online learning materials effectively. Familiarize yourself with common assistive technologies and design your courses to be compatible with them. Test your course materials with these technologies, ideally with students who use them, to ensure full functionality. This proactive approach eliminates potential accessibility barriers before your course launches. It's also wise to provide clear instructions on how to use relevant assistive technologies within your learning management system, supporting students in accessing the tools that meet their unique needs. Beyond technological solutions, fostering an inclusive online community is paramount. Encourage respectful communication and collaboration, creating a space where students from diverse backgrounds feel welcome and supported. Explicitly address accessibility in your course syllabus, clarifying your commitment to inclusivity and outlining the support available. Regularly check in with students to identify and address any accessibility barriers they encounter. Solicit feedback from students throughout the course, ensuring that their perspectives inform your teaching practices and future course design. This continuous feedback loop guarantees that the course remains inclusive and accessible, fostering a positive learning environment.

Finally, remember that promoting accessibility is an ongoing process, not a one-time task. The needs of your learners may evolve, and new technologies and pedagogical approaches continually emerge. Staying abreast of these developments is crucial to maintaining accessible and inclusive online courses. This might involve attending workshops, engaging with online communities dedicated to accessibility, and participating in professional development opportunities focused on inclusive teaching practices. By committing to these strategies, you will ensure your online courses are truly accessible to all learners, enhancing their educational experience and maximizing their potential for success.

6 Technology Integration in Higher Ed

Selecting the right technology isn't about choosing the flashiest tools; it's about aligning technology with pedagogical goals. Consider your learning objectives. What specific skills or knowledge do you want students to acquire? Once you've clearly defined these, research technologies that directly support those aims. For instance, if collaboration is key, explore platforms that facilitate group projects and real-time communication. If visual learning is dominant, consider incorporating video lectures or interactive simulations. Avoid falling into the trap of adopting technology simply because it's popular; ensure its functionality matches your specific needs and enhances, not hinders, the learning process. The effectiveness of your technology choices will be directly proportional to how well they support your teaching philosophy and student learning styles.

Effective integration means more than just uploading lecture slides to a learning management system (LMS). It requires thoughtful planning and implementation, beginning with faculty training. Investing in comprehensive training ensures that instructors understand the technology's capabilities and how to integrate them seamlessly into their courses. Training shouldn't be a one-off event; ongoing support and professional development opportunities are crucial for sustained effectiveness. Remember, effective technology integration necessitates a culture of innovation and collaboration among faculty, fostering a collective understanding and sharing of best practices. This collaborative effort can then be leveraged to create robust and effective technology-rich learning environments.

Data analytics offer powerful insights into the effectiveness of your technology integration strategies. By analyzing student engagement data—such as time spent on different modules, completion rates, and quiz scores—you can identify areas for improvement. For example, if students consistently struggle with a particular module, you can refine the content, add supplementary materials, or adjust the pacing. Furthermore, tracking student interactions with various technologies provides invaluable feedback on their effectiveness. This data-driven approach allows for continuous improvement and refinement of your online courses, ensuring that your technology choices are continuously optimized for maximum learning impact. This continuous cycle of data analysis and iterative improvements is critical to maximize learning outcomes and resource use.

Beyond analyzing individual student performance, data analytics can reveal broader trends. Are certain learning styles or demographics better served by particular technologies? Are specific tools consistently underutilized? Understanding these patterns allows you to make more informed decisions about future technology investments and course design. By regularly reviewing this data, you can adapt your teaching strategies and resource allocation to better meet the needs of your student population, creating a more equitable and effective learning experience for all. Remember, data-driven decisions lead to greater personalization, creating positive outcomes for students and educators alike.

Integrating technology effectively in higher education isn't just about adding tools; it's about transforming the learning experience. Consider how technology can foster collaboration and communication. Tools like online forums, group projects, and virtual office hours promote a sense of community among students, even when geographically dispersed. Think about how technology can personalize the learning path. Adaptive learning platforms can adjust the difficulty and pace of instruction based on individual student needs, ensuring that each student receives the optimal level of support and challenge. The ultimate goal should always be student success and the enhanced educational experience made possible by thoughtfully implementing technology.

Implementing new technologies requires careful consideration of accessibility. Ensure that all chosen technologies are compatible with assistive technologies used by students with disabilities. This includes providing alternative text for images, captions for videos, and keyboard navigation for all interactive elements. Accessibility should be considered not an afterthought but a core element of the design process from the outset. By prioritizing accessibility, you create an inclusive learning environment where all students can fully participate and succeed. Inclusive design creates not only an equitable but also a more enriching educational experience for all students. Remember, making an inclusive environment shows a commitment to improving education for everyone.

6.1 Choosing the Right Technologies

The selection of appropriate technology is paramount for successful e-learning. It's not simply about choosing the latest, shiniest tools; it's about aligning technology with pedagogical goals and institutional resources. A poorly chosen Learning Management System (LMS), for example, can hinder collaboration and severely impact the overall learning experience, leading to frustration for both instructors and students. Conversely, a well-integrated system can streamline administrative tasks, boost student engagement, and provide valuable data for continuous improvement.

Consider your institution's specific needs. A small college with limited IT support may find a cloud-based LMS with robust customer support more beneficial than a complex, on-premise system requiring extensive technical expertise. Larger universities, however, with dedicated IT departments and greater financial resources might opt for a more customizable, on-premise solution, allowing for greater control and integration with existing systems. This detailed assessment ensures alignment between technological capabilities and practical limitations.

Accessibility is a critical factor. The chosen technologies must comply with accessibility standards, such as WCAG (Web Content Accessibility Guidelines), ensuring that all students, regardless of their abilities, can fully participate in the online learning environment. This includes features like screen readers compatibility, keyboard navigation, captioning for videos, and alternative text for images. Ignoring accessibility not only compromises inclusivity but can also expose the institution to legal challenges. Therefore, a thorough evaluation of accessibility features is crucial before committing to any technology.

Beyond the LMS, consider the specific needs of individual courses. For example, a science course might require specialized software for simulations or data analysis, while a humanities course might benefit from collaborative writing tools or video conferencing software. Integration between different platforms is crucial for a seamless learning experience, avoiding a fragmented and confusing environment for students navigating multiple systems. Efficient integration ensures a positive learning environment.

Cost is another significant consideration. While some technologies offer free versions or open-source alternatives, they may lack essential features or require significant time and expertise for setup and maintenance. Therefore, a realistic budget is necessary, weighing the costs of software licenses, hardware, technical support, and ongoing training against the potential benefits for teaching and learning. Free options often lack essential features.

Data analytics can provide crucial insights into the effectiveness of chosen technologies. Tracking student engagement metrics, such as time spent on modules, completion rates, and participation in discussions, offers valuable feedback for refining teaching strategies and improving the learning design. Analyzing usage patterns of various tools and platforms can identify areas for optimization, ensuring resources are deployed efficiently and effectively. This data-driven approach maximizes the value of technology investments.

Finally, continuous evaluation is essential. Technology is constantly evolving, and what works today might become obsolete tomorrow. Regular reviews of chosen technologies, including feedback from instructors and students, ensures the system remains aligned with evolving pedagogical approaches and technological advancements. This cyclical approach ensures that the selected systems continuously meet the ever-changing requirements of a dynamic educational landscape. Therefore, the initial selection of technology is not a one-time decision; it is the starting point for an ongoing process of refinement and adaptation.

6.2 Data Analytics for Improvement

Harnessing the power of data analytics is crucial for refining your e-learning strategy and maximizing its impact. By systematically collecting and analyzing data, you can gain valuable insights into student performance, course effectiveness, and overall program success. This empowers you to make data-driven decisions that lead to continuous improvement.

Effective data analytics begins with identifying key performance indicators (KPIs). What metrics truly matter for your institution? Consider factors like student completion rates, average time spent on modules, assignment scores, participation in discussions, and overall student satisfaction. These metrics offer a holistic view of your program's health. Defining these KPIs early on is critical for a targeted analysis, allowing you to focus your resources and time efficiently. Without clear goals, data collection can become a wasteful exercise.

Once you've established your KPIs, the next step involves selecting the right tools for data collection and analysis. Learning management systems (LMS) often provide built-in analytics dashboards, offering a wealth of information about student activity. However, more sophisticated analysis may require specialized analytics software or even the expertise of a data analyst. There is a range of solutions available, from simple spreadsheet programs to advanced statistical packages. The appropriate choice depends on your institution's technical capabilities and the complexity of your analysis needs. Careful consideration should be given to the level of detail you need and your budget.

Analyzing the collected data is where the real insights emerge. Look for patterns and trends in student performance. Are there particular modules that students struggle with? Are there specific learning styles that correlate with higher success rates? Identifying such trends allows for targeted interventions, such as revising challenging modules, adjusting teaching methodologies, or offering additional support to struggling students. Correlation doesn't always equal causation, but recognizing trends can provide strong clues for improvement.

Beyond individual student performance, you can use data analytics to evaluate the overall effectiveness of your e-learning program. Compare student outcomes in online courses with those in traditional face-to-face classes, if applicable. Analyze the impact of different teaching strategies, technologies, and learning materials on student engagement and achievement. This larger-scale analysis allows for strategic improvements across your e-learning program. This is particularly useful for demonstrating the impact of your program and justifying your choices to stakeholders.

Data analytics isn't just about identifying problems; it's also about celebrating successes. Highlighting positive trends and outcomes can boost morale and encourage ongoing improvement. Showcase successful teaching strategies, effective learning materials, and student achievements. Public recognition of achievements, supported by data, motivates both educators and learners. This positive reinforcement further encourages a culture of continuous improvement.

Finally, remember that data analytics is an ongoing process, not a one-time event. Regularly monitor your KPIs, analyze your data, and adapt your strategies accordingly. The educational landscape is constantly evolving, and your e-learning program should evolve with it. Continuous monitoring allows you to adapt to changing student needs and technological advancements. By making data-driven decisions, you'll be well-positioned to create an effective and continuously improving e-learning environment. The iterative nature of this process is essential for sustained success in this ever-changing field. Your dedication to refinement and improvement will directly translate to the success of your students and the overall positive impact of your program.

7 Future Trends in E-Learning

The landscape of e-learning is poised for a dramatic transformation. We're moving beyond simple online courses to immersive, personalized learning environments powered by cutting-edge technologies. Imagine virtual reality simulations allowing students to dissect a frog's heart without harming a living creature, or augmented reality overlays providing real-time feedback on a student's experiment, guiding them towards better understanding through immediate visual cues. These aren't just futuristic fantasies; they are technologies rapidly maturing and becoming more accessible.

Expect a significant increase in the use of artificial intelligence, not just for grading and administrative tasks, but for truly personalized learning experiences. AI-powered tutors can adapt to individual learning styles and paces, providing customized content and support. Imagine an intelligent system that recognizes a student's struggles with a particular concept and instantly provides tailored resources, from interactive exercises to relevant videos, all within the flow of their learning journey. This level of individualized attention is currently impossible to achieve at scale with human instructors alone.

The rise of microlearning and personalized learning pathways are directly tied to the capabilities of AI. These technologies are enabling the creation of modular learning content which can be combined to create customized courses that cater specifically to a student's background, goals, and learning style. Instead of a rigid curriculum, students will have the autonomy to chart their own learning paths, choosing the resources and topics that best meet their needs. This empowers learners to take ownership of their education, fostering a deeper sense of engagement and accomplishment.

However, the ethical considerations surrounding AI in education are paramount. Concerns about bias in algorithms, data privacy, and the potential displacement of human instructors must be carefully addressed. Open discussions and robust regulations are necessary to ensure that the integration of AI is both effective and ethical, maximizing its benefits while mitigating its risks. A balanced approach, combining the strengths of human interaction with the capabilities of AI, will be crucial for a positive impact.

Furthermore, the metaverse and its potential applications in e-learning are beginning to show their promise. Imagine collaborative virtual classrooms where students can interact with each other and their instructors in a three-dimensional environment, fostering a stronger sense of community. The immersive nature of the metaverse could revolutionize fields like medicine, engineering, and art, allowing students to interact with complex simulations and virtual environments in ways that were previously impossible. The possibilities for interactive learning experiences are virtually limitless.

The future of e-learning also hinges on the seamless integration of various technologies. We're not just talking about combining VR and AI; it's about creating ecosystems where these technologies work together flawlessly to enhance the entire learning experience. This necessitates a shift in how we approach educational technology—from a focus on individual tools towards a holistic, integrated approach that prioritizes user experience and learning outcomes.

Moreover, the continued growth of open educational resources (OER) is another significant trend. OER are freely accessible educational materials, and their increasing availability is democratizing education, making high-quality learning resources available to a broader audience, regardless of socioeconomic background or geographical location. The collaborative nature of OER development further strengthens its potential, drawing upon the expertise and creativity of a global community of educators and learners.

However, the proliferation of OER also presents challenges. Ensuring the quality and accuracy of these resources, while maintaining ethical guidelines around copyright and attribution, requires careful consideration. The development of robust quality assurance mechanisms and effective community governance models will be essential for managing and expanding the use of OER in a sustainable and responsible manner. Balancing the freedom and accessibility of OER with the need for reliable and trustworthy information will be a crucial task for the future.

Finally, the ongoing evolution of e-learning necessitates a shift in mindset, both for educators and learners. It's no longer enough to simply adapt traditional teaching methods to an online format. We must embrace innovative pedagogical approaches, design learning experiences that are truly engaging and relevant, and foster a culture of lifelong learning. This requires investment in professional development for educators, enabling them to develop the skills and knowledge necessary to effectively utilize emerging technologies and create dynamic, impactful learning environments. Ultimately, the future of e-learning hinges on our ability to adapt and innovate, ensuring that technology serves the ultimate goal of empowering learners and enhancing the quality of education. The success of future e-learning depends on this collaborative approach, combining technological advancements with a deep understanding of effective pedagogy and a commitment to equitable access for all.

7.1 Emerging Technologies

Virtual Reality (VR) and Augmented Reality (AR) are poised to revolutionize e-learning. Imagine students dissecting a virtual heart in a biology class, or walking through ancient Rome during a history lesson, all from the comfort of their homes. These immersive technologies offer unparalleled opportunities for experiential learning, surpassing the limitations of traditional textbook learning. The realistic simulations fostered by VR and AR can significantly enhance engagement and comprehension, especially for subjects requiring hands-on experience. The potential for cost savings is substantial; expensive field trips or laboratory equipment become less necessary, making quality education more accessible. Furthermore, the interactive nature of VR and AR allows for personalized learning experiences tailored to individual student needs and paces. The development of user-friendly VR and AR platforms specifically designed for educational purposes is key to realizing their full potential.

Blockchain technology, known for its secure and transparent nature, offers intriguing possibilities for e-learning. Its decentralized nature ensures data integrity, protecting student records and academic credentials from tampering or loss. This feature is critical for building trust and maintaining the credibility of online degrees and certifications. Furthermore, blockchain can facilitate the creation of secure digital learning assets, such as certificates and badges, that are easily verifiable and shareable. This is particularly beneficial in the context of lifelong learning, where individuals accumulate credentials from various sources. By enabling seamless verification of achievements, blockchain can streamline the process of academic record-keeping and enhance the transparency and credibility of online learning programs. However, careful consideration must be given to the potential challenges associated with implementing this complex technology within an educational setting. Understanding the technical requirements and establishing suitable governance models are paramount.

The increasing prevalence of big data in education provides valuable insights into student learning patterns and preferences. By analyzing student interactions with learning materials and assessments, educators can identify areas where interventions might be necessary. This data-driven approach empowers educators to personalize learning paths, target specific learning challenges, and enhance the overall effectiveness of online instruction. For instance, analyzing student performance on quizzes can reveal concepts requiring additional explanation, prompting teachers to adjust their teaching strategies accordingly. This allows for a more dynamic and responsive educational experience, addressing students' individual needs in real-time. This sophisticated analysis, however, necessitates careful consideration of data privacy and security. Protecting student information is essential, and robust data governance strategies must be in place to ensure compliance with relevant regulations.

The Internet of Things (IoT) is gradually transforming the landscape of education. Smart devices and sensors integrated into learning environments can capture real-time data on student engagement and learning progress. This enables educators to monitor students' activity, identify areas of difficulty, and provide timely feedback. For example, sensors tracking students' eye movements during online lectures might reveal disengagement, prompting intervention to ensure comprehension. While this provides invaluable insights, potential privacy issues must be proactively addressed. Transparency and explicit consent from students regarding data collection and usage are crucial. Furthermore, the integration of IoT devices into e-learning requires careful planning and integration into the existing technological infrastructure. The focus must remain on enhancing learning outcomes and not simply on technology adoption for its own sake. The ethical considerations and potential biases in data interpretation also need to be carefully considered and mitigated.

Adaptive learning platforms powered by sophisticated algorithms represent a significant advancement in personalized education. These platforms analyze student performance in real time, adapting the learning content and pace to each individual's needs. This ensures that students are always challenged at an appropriate level, preventing boredom from being underserved, and frustration from being overwhelmed. For example, if a student consistently struggles with a particular concept, the platform might provide additional practice exercises or supplementary materials tailored to their specific learning style. The resulting individualized learning experience is particularly beneficial for diverse student populations, catering to varied learning paces and styles. However, these platforms require careful calibration to avoid perpetuating existing biases present in the data they are trained on. Attention must also be paid to the limitations of algorithm-based assessments and the importance of maintaining human interaction and mentorship within the learning process. The selection and careful evaluation of such platforms is therefore paramount.

7.2 The Future of Online Learning

The metaverse offers unprecedented opportunities for immersive learning. Imagine students collaborating on a virtual archaeological dig, exploring a 3D model of the human heart, or attending a live concert from a historical period – all within a shared digital space. This isn't science fiction; prototypes are already being tested, and the potential to create truly engaging and memorable learning experiences is immense. The challenge lies in bridging the gap between current technology and pedagogical best practices, ensuring that these immersive environments enhance, rather than simply replace, effective teaching strategies.
Personalized learning will reach new heights. Artificial intelligence, coupled with advanced data analytics, will allow educators to tailor learning pathways with remarkable precision. Imagine a system that continuously assesses a student's understanding, adapting the difficulty and content in real-time to optimize their learning trajectory. This will move beyond simple adaptive quizzes; it could involve personalized feedback, customized assignments, and even the creation of unique learning modules based on individual strengths and weaknesses. The key is responsible implementation, ensuring that personalization doesn't lead to isolation but instead fosters a sense of individual support within a collaborative learning community.
Micro-credentialing and skills-based learning will become increasingly important. In a rapidly changing job market, individuals need

to continuously update their skill sets. Micro-credentials, which verify specific competencies, will offer a flexible and accessible way to acquire new skills. Online platforms will offer a wide array of short, focused courses, allowing learners to acquire specific skills relevant to their career goals. This allows for greater career mobility and adaptability, making online education a vital tool for lifelong learning and professional development. The key here is ensuring the quality and recognition of these micro-credentials, working with industry partners to establish clear standards and pathways for career advancement.

The role of the educator will evolve significantly. Instead of being primarily content deliverers, instructors will become facilitators, mentors, and curriculum designers. They will leverage technology to guide students through personalized learning journeys, fostering collaboration and critical thinking. This necessitates a shift in teacher training, equipping educators with the skills to design and implement innovative online curricula and effectively manage diverse learning environments. The focus will be less on lecturing and more on guiding, coaching, and providing individualized support.

Accessibility will be paramount. The future of online learning must ensure that everyone has equal access to high-quality education, regardless of their location, socioeconomic status, or disability. This requires a multi-faceted approach, encompassing accessible technology, inclusive curriculum design, and supportive learning environments that cater to diverse learning styles and needs. The development of universal design principles for online learning will be crucial, creating learning experiences that are inherently accessible to all.

The integration of online and offline learning will become the norm. The most effective approach may not be entirely online or entirely face-to-face but a blended learning model that leverages the strengths of both. This blended approach allows for flexible learning, individualized attention, and a sense of community. This requires careful consideration of how to integrate these two modes effectively, creating a seamless and engaging learning experience that takes advantage of both the flexibility of online learning and the benefits of face-to-face interaction. Careful planning and communication are key to avoiding inconsistencies and ensuring a coherent learning journey.

Ultimately, the future of online learning is bright, full of potential to revolutionize how we learn and grow. However, realizing this potential requires a proactive and thoughtful approach, one that prioritizes pedagogical excellence, accessibility, and the ongoing professional development of educators. By embracing innovation and focusing on the needs of learners, we can create a truly transformative online learning ecosystem. This will not only enhance educational opportunities but also empower individuals to thrive in an increasingly complex and rapidly changing world. The challenge lies in effectively navigating these developments, ensuring that technology enhances the learning process, rather than dictating it.

7.3 AI in Education

Artificial intelligence (AI) offers transformative potential for education, moving beyond simple automation to create truly personalized and adaptive learning experiences. Imagine a system that instantly tailors learning materials to individual student needs, adjusting difficulty and pace based on real-time progress. This is not science fiction; AI-powered platforms are already doing this, providing immediate feedback and suggesting alternative learning paths to ensure comprehension. This targeted approach drastically improves learning outcomes, particularly for students who struggle in traditional classroom settings or require additional support.

The implementation of AI in education isn't simply about replacing teachers; rather, it's about augmenting their capabilities. AI can automate tedious administrative tasks like grading and scheduling, freeing up educators to focus on what truly matters: individualized student interaction and mentorship. Consider the time saved by automatically grading objective assessments, allowing instructors to dedicate more time to providing insightful feedback on essays and projects. This shift in focus allows for deeper engagement with students and more effective personalized instruction.

Furthermore, AI offers unprecedented opportunities for creating engaging and interactive learning environments. Imagine immersive simulations that allow students to experience historical events first-hand or practice complex scientific procedures in a safe and controlled virtual setting. AI-powered chatbots can offer immediate support and guidance to students struggling with concepts, answering questions and providing supplementary learning resources. This always-available support system addresses the common challenges of asynchronous learning, creating a more supportive and accessible educational experience.

However, ethical considerations are paramount. Concerns around data privacy, algorithmic bias, and the potential displacement of human educators require careful attention. Transparency in how AI systems operate is critical, and ongoing research into mitigating bias in algorithms is essential to ensure fair and equitable access to educational opportunities for all students. The role of the human teacher remains indispensable, providing the crucial element of human connection and fostering a sense of community that cannot be replicated by technology alone. Therefore, a thoughtful and strategic implementation of AI must focus on augmenting, not replacing, the human element in education.

AI's impact extends beyond the immediate learning process. Data analysis powered by AI can provide educators with valuable insights into student performance and learning patterns, identifying areas where intervention is needed and informing improvements to curriculum and teaching strategies. Imagine using this data to pinpoint specific learning challenges experienced by a group of students, allowing for proactive adjustments to instructional materials and teaching techniques to better support their needs. This data-driven approach promotes continuous improvement and allows for more evidence-based decision-making in the design and delivery of educational programs. We must strategically adopt and integrate AI to unlock its educational potential while actively addressing the associated challenges.

The effective integration of AI in education necessitates careful planning and consideration. It requires ongoing professional development for educators to understand and utilize AI tools effectively, ensuring they can integrate these technologies seamlessly into their teaching practices. Moreover, the digital literacy of students must also be considered and adequately addressed; students must be equipped with the skills to navigate and critically evaluate information presented by AI-powered systems. The successful adoption of AI in education will not simply involve the introduction of new technologies, but also require a concerted effort to upskill both educators and students, ensuring everyone is well-prepared for this transformative shift in the educational landscape. The focus should always remain on creating a supportive and engaging learning environment that prioritizes student success and empowers educators.

www.ingramcontent.com/pod-product-compliance
Lightning Source LLC
LaVergne TN
LVHW022013200926
848481LV00007B/2375
9798349285622